TOP LEFT AND CLOCKWISE: YELLOW-BELLIED GLIDER; SUGAR-GLIDER; MALE RED KANGAROO; KOALA; ORANGE-EYED TREE-FROG; LITTLE PENGUIN; AUSTRALIAN PELICAN; DINGO

ROCKY COASTLINE TYPICAL OF SOUTHERN AUSTRALIA

AUSTRALIA: HOME TO WILDLIFE

Australia was once part of a great southern super-continent called Gondwana. Rifting and spreading of the sea-floor allowed continents which today are India, Antarctica and Australia to drift away. In the last 50 million years Australia has moved away from Antarctica, carrying animals which have continued to evolve in isolation.

As the continent drifted northwards, climates became hotter and drier and rainforests which had covered vast areas changed to savannahs or shrank to tiny coastal remnants. Many of the animals and plants which exist in modern Australia have adapted to cope with arid conditions.

As rainfall decreases from east to west across the continent, woodlands and scrublands shade into plains covered with native grasses and spinifex. Much of the centre and north-west of Australia is desert. In the tropical north, annual monsoon rains create Wet and Dry seasons. In the extreme South-West and South-East of the continent, abundant rain falls mainly in winter.

There are many climatic regions in this great landmass, but the theme for most of Australia is regular or seasonal aridity. Overall, it is the driest continent on the planet. Over much of the continent, wildlife is most easily found near water, though some species are adapted to go for long periods without drinking. Breeding seasons are timed to coincide with rain or follow quickly after rain falls.

Inland waterways and coastal areas have their own wildlife, while the coral reefs of the northern coastline offer a wonderland of marine life.

EUCALYPTS ARE COMMON IN DRIER FORESTS

TOP TO BOTTOM: RAINFOREST REQUIRES REGULAR, ABUNDANT RAINFALL; SANDY DESERT WHERE MANY ANIMALS ARE NOCTURNAL; TROPICAL FLOODPLAIN INUNDATED BY MONSOON RAINS; ARID RANGES OF INTERIOR ARE HOME TO MANY ANIMALS; GREAT BARRIER REEF AND ITS ISLANDS: A WONDERLAND

KOALA

This popular symbol of Australia eats only the leaves of the eucalypt trees in which it lives. Each leaf is carefully inspected before it is eaten. The diet is low in energy and a Koala spends up to 19 hours out of each 24 asleep and most of the rest of the time feeding.

Two "thumbs" on each hand and long, strong limbs help make the tailless Koala a good climber. Sleep periods are spent in tree forks.

The female bears a tiny young one 35 days after mating. The baby climbs into her pouch, where it matures until it has fur and can cling to its mother's back. It will stay with her for another year before venturing off on its own.

Koalas from the southern parts of Australia are larger than those from Queensland. Male koalas are up to 50 per cent heavier than females.

Fossil evidence shows there were Koala-like animals in Australia 14 million years ago. Today, the Koala's closest living relative is the wombat. The continued existence of wild Koalas depends on a network of suitable eucalypt habitats.

ABOVE: THE KOALA EATS ONLY EUCALYPT LEAVES
OPPOSITE: A YOUNG KOALA CLINGS TO ITS MOTHER'S BACK

THE KOALA IS SUPERBLY ADAPTED TO LIFE IN THE TREETOPS

TOP LEFT AND CLOCKWISE: LEAVES ARE CAREFULLY INSPECTED BEFORE BEING EATEN; HANDS AND FEET GRASP TREE FIRMLY; THERE IS A LONG PERIOD OF MATERNAL CARE; THE KOALA HAS NO VISIBLE TAIL AND LONG LIMBS

KOALAS PREFER TO EAT JUST A FEW SPECIES OF EUCALYPTS, INCLUDING MANNA GUM AND FOREST RED GUM

EVEN WHEN IT IS ASLEEP, THERE IS LITTLE CHANCE OF A KOALA FALLING. NOTE THE TWO 'THUMBS' ON EACH HAND

PARMA WALLABY

GOODFELLOW'S TREE-KANGAROO, A NEW GUINEA SPECIES

RUFOUS BETTONG

KANGAROOS

In Australia, kangaroos and their relations occupy the niches filled in other continents by deer, antelopes and other grazing and browsing animals. The macropod (large-footed) group consists of kangaroos, wallabies, tree-kangaroos, rat-kangaroos and the tiny Musky Rat-Kangaroo. Members of the kangaroo family usually have big, muscular hindlegs and long tails. When moving fast, they rise on their hindlegs and hop, using their tails to balance. All females of the group have forward-opening pouches and the single joey, like the young of all marsupials, is born undeveloped and matures in the pouch.

Some sorts of kangaroos have held their own in the face of European settlement. The small potoroos, bettongs and pademelons have suffered from habitat destruction and the depredations of feral animals such as foxes and cats. The Quokka and some other small species have survived on coastal islands out of reach of predators. Tree-kangaroos live in the restricted rainforests of eastern Cape York.

CLOCKWISE FROM TOP LEFT: RED-LEGGED PADEMELON; RED KANGAROO; RED-NECKED PADEMELON; TASMANIAN PADEMELON; WESTERN GREY KANGAROO; BENNETT'S WALLABY
OPPOSITE: EASTERN GREY KANGAROO FEMALE WITH JOEY IN POUCH

TOP LEFT AND CLOCKWISE: QUOKKA; SWAMP WALLABY; BRUSH-TAILED ROCK WALLABY; RED-NECKED WALLABY; YELLOW-FOOTED ROCK WALLABY; COMMON WALLAROO; PRETTY-FACE WALLABY AND JOEY; ANTILOPINE WALLAROO; AGILE WALLABY. OPPOSITE: MALE RED KANGAROO HAS MUSCULAR FOREARMS; INSET LEFT AND RIGHT: THE FEMALE RED KANGAROO (FAR RIGHT) MAY BE GREY IN COLOUR; HOPPING RED KANGAROO USES TAIL AS A COUNTER-BALANCE

PLATYPUS

The Platypus and the echidna are mammals. They are warm-blooded, possess hair and feed their young on milk. However, they are remarkable in that they lay soft-shelled eggs and the young ones lap milk from their mother's abdomen as she has no nipples.

The Platypus is found in waterways, where it swims with its webbed forefeet with the hindfeet folded back against the broad, flat tail when not being used for steering or as brakes. The eyes, nostrils and earholes are kept shut underwater. The sensitive, rubbery bill picks up signals from food animals, which are grabbed and stored in cheek-pouches until the Platypus surfaces to eat them. A baby Platypus has milk teeth, but when these are shed they are not replaced. Adults grind up food between the tongue and horny plates on the jaws.

Home is a burrow in the creekbank. The two eggs are incubated for one to two weeks, then the young fed on milk for four to five months.

On each ankle, the male Platypus has a hollow spur, which serves as outlet for a venom gland in each groin.

THE AQUATIC PLATYPUS HAS WEBBED FEET, A THICK PELT AND CAN CLOSE EYES, EARS AND NOSTRILS WHEN SWIMMING

SHORT-BEAKED ECHIDNA

Like the Platypus, the echidna is a monotreme, a mammal whose young hatch from eggs and are then fed on milk. Australia's Short-beaked Echidna feeds on ants and termites, which it obtains by pulling a nest apart with strong-clawed forearms, then inserting its long snout and tongue. Insects adhere to the saliva-covered tongue, are drawn back into the mouth and crushed between a horny pad at the back of the tongue and another pad on the palate.

Echidnas are found all over Australia and Tasmania. In colder areas, they have thick fur between their sharp spines and may hibernate when temperatures fall drastically.

A frightened echidna will dig rapidly down into soft ground, leaving only the tips of the spines visible.

Two weeks after mating, the female lays an egg into the pouch on her belly. It hatches after 10 days and the young one remains in the pouch for another three months. When it develops spines, it is left in a "den" while the mother seeks food.

THE ECHIDNA HAS PROTECTIVE SPINES AND STRONG DIGGING CLAWS

DINGO

The oldest Dingo remains found in Australia are of animals which died less than 4,000 years ago. The origins of this wild dog seem to have been in south-east Asia and its arrival in Australia may have signalled the end of a less effective predator, the Tasmanian Wolf (Thylacine), on the mainland. Some Dingos lived in association with the Aboriginal people, but after European settlement their place in camp was often taken by domestic breeds of dog. The name used by the Port Jackson Aborigines to refer to their camp dingoes may come from "tingo", meaning "tame".

The majority of Dingos are yellow-ginger in colour, but some are black and a few are white. The stronghold of the Dingo today is the forested country between the Great Dividing Range and the eastern coast. Where small animals such as rabbits are the prey, Dingos live alone or in small groups. Packs are formed to hunt larger prey such as kangaroos. Domestic livestock make up less than 2 per cent of the diet of Dingos in eastern Australia. The female Dingo comes into season only once each year. Dingos and domestic dogs will interbreed.

A DINGO DRINKS AT A COASTAL FRESHWATER STREAM
OPPOSITE: THE DINGO'S HOME RANGE MAY BE UP TO 20 KMS ACROSS

COMMON WOMBAT

DRIVE CAREFULLY – WOMBATS CROSS ROAD HERE

WOMBATS

There are three species of wombat, the Common, the Southern Hairy-nosed and the extremely rare Northern Hairy-nosed, which now numbers perhaps 80 individuals found only in a small area in central Queensland.

Wombats live in burrows and a wombat warren may have up to 12 entrances, with individual tunnels extending up to 30 metres. The burrow is the centre of wombat existence, a secure refuge which protects against heat, cold, dehydration and even bushfires. Several wombats may co-exist in a warren, with little interaction between individuals except at mating time. The wombat emerges to eat vegetation, with teeth which have no roots and continue to grow during the wombat's lifetime.

Wombats are protected by law in most parts of Australia. They have the best-developed brains of any marsupial and are remarkably agile for their shape, being able to run at 40 kph for a short distance. A Common Wombat may live for up to 20 years.

THE RARE NORTHERN HAIRY-NOSED WOMBAT
OPPOSITE: THE WOMBAT HAS THE BEST-DEVELOPED BRAIN OF ANY MARSUPIAL

WOMBATS EAT GRASS, LEAVES AND ROOTS

QUOLLS

Quolls are cat-sized, spotted carnivorous marsupials, whose diet ranges from small mammals and birds to insects. They are athletic, agile and active climbers. The Eastern Quoll, the most terrestrial of the four species, now survives only in Tasmania, though it once ranged over most of south-eastern Australia. The Western Quoll has declined to a range in the south-west of Western Australia. The Spotted-tailed Quoll is a large, aggressive hunter now limited to coastal eastern Australia. The small Northern Quoll is found across the north of the continent.

Females quolls have restricted home ranges, while males move about more widely. They may share dens but otherwise do not interact much except in the breeding season. A female quoll may give birth to large numbers of young ones but has only six nipples for them to attach to. Once the young are too big for the pouch, the female leaves them in a nest while she hunts. She may carry them on her back when moving them from one den to another. Young quolls are very playful.

SPOTTED-TAILED QUOLL

TOP: NORTHERN QUOLL BOTTOM: EASTERN QUOLL

SMALL MAMMALS

Australia contains representatives of three groups of mammals. Monotremes lay eggs and suckle their hatchlings on milk oozing from pores on the abdomen. Marsupials and placental mammals bear young which are suckled on milk ducted through nipples. However, a baby marsupial is born in an embryonic condition and develops for some time in the mother's pouch. A baby placental mammal is nourished within the mother's body until comparatively late in its development. A small furry animal in Australia may be a marsupial, or a native placental mammal, or an introduced placental mammal.

Phascogales, antechinus and dunnarts are small carnivorous marsupials. Males of the first two species may be so stressed at the breeding season that they die after mating. Bandicoots are ground-dwelling marsupials which eat spiders, earthworms and insects. The hopping-mice are placental mammals which survive in arid areas by living in burrows and obtaining water from their diet of seeds, plants and occasional insects.

TOP LEFT AND CLOCKWISE: BRUSH-TAILED PHASCOGALE; BARRED BANDICOOT; FAT-TAILED DUNNART; SPINIFEX HOPPING-MOUSE
OPPOSITE: YELLOW-FOOTED ANTECHINUS

BILBY

Bilbies are bandicoots which dig burrows and have long well-furred tails, long silky fur and long rabbit-like ears. One of the two species, the Lesser Bilby, is probably extinct.

Once widespread in Australia's savannah woodland and shrub grassland, the Bilby is now confined to the arid lands of the Centre with a few scattered outposts in northern Western Australia.

This beautiful animal remains in its burrow during daylight and feeds on seeds, insects and bulbs at night, digging for food with strong forelimbs. Two young are reared in the pouch for about 10 weeks, then suckled in the burrow for another two weeks.

Land degradation by domestic livestock and fire, competition from rabbits and predation by foxes and cats have all contributed to the Bilby's decline. Captive breeding programmes are researching the biology and behaviour of this beautiful marsupial and it may be possible to re-establish it over some of its former range.

ABOVE: AUSTRALIA'S ARIDLANDS FORM A FRAGILE HABITAT FOR THE BILBY
LEFT: THE BILBY EATS SEED AND SMALL CREATURES DUG FROM THE SOIL

TASMANIAN DEVIL

While the Tasmanian Devil, with its powerful jaws and sharp teeth, looks fierce, it is not a hunter but a scavenger. A number of Tasmanian Devils may gather at food such as a road-kill and although physical clashes are rare there will be much open-jawed threatening and loud growling and yelling. Females are smaller than males. Three or four young are born in April or May and develop in the pouch for about 15 weeks. After that, the mother feeds them for another three or four months, leaving them in a den until they are old enough to follow her.

The Tasmanian Devil was present on the mainland of Australia as recently as 600 years ago, but now survives only in Tasmania.

THE TASMANIAN DEVIL IS A SHORT-LEGGED, STRONG MARSUPIAL SCAVENGER

YOUNG TASMANIAN DEVILS ARE VERY PLAYFUL

GHOST BAT

BATS

Bats are placental mammals and the only mammals capable of sustained flight. A bat's wing consists of thin membrane whose leading edge joins shoulder and thumb and whose trailing edge joins the ends of the fingers, then extends to the ankle and in varying degrees to the tail. Australia has many species of small insect-eating bats and one larger Ghost Bat, which is a predator on small animals, including other bats. All use echo-location as well as sight to navigate and hunt and many become torpid in cold weather. The comparatively large fruit-bats or flying-foxes do not use echo-location, have no webbing between their hindlegs and feed on plant juices and fruit. Intelligent and appealing animals, they inhabit warmer parts of the continent and may fly long distances to feed.

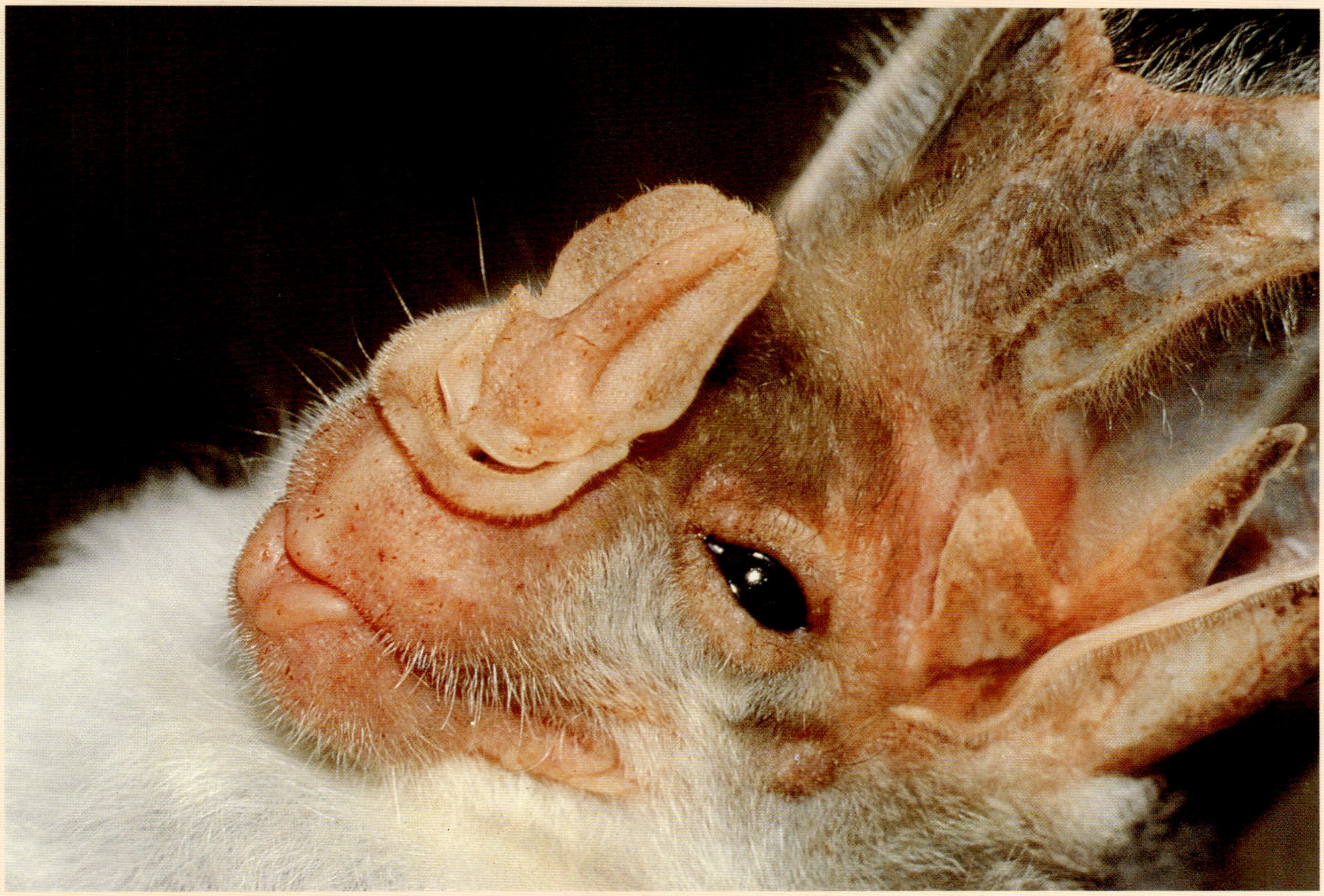

THE GHOST BAT'S LARGE EARS AND FACIAL WATTLES ASSIST ITS USE OF ECHO-LOCATION
LEFT: THE GREY-HEADED FRUIT-BAT RESTS UPSIDE-DOWN DURING DAYTIME

POSSUMS

Possums are a group of marsupials with long, flexible tails and forward-opening pouches. They walk or run rather than hop and most are agile climbers, living in trees or amongst rocks.

Brushtail possums have bushy tails which are only moderately useful for clinging onto branches. Ringtails and cuscuses have coiled tails which are very useful "fifth limbs".

Gliders possess a fold of skin on each side of their body, joining forelimb to hindlimb, which enables them to plane from tree to tree. The rare Leadbeater's Possum, although without gliding membranes, strongly resembles the Sugar Glider. The mouse-sized pygmy-possums have prehensile tails and eat insects and nectar. Sap, gum and manna from eucalypts and wattles are also eaten by possums and gliders.

The ground-dwelling Mountain Pygmy-possum survives winter in tunnels under the snow of the Australian Alps.

A few possum species have adapted well to human habitation and are to be found in parks and gardens in towns and cities.

TOP LEFT AND CLOCKWISE: FEATHERTAIL GLIDER; MOUNTAIN PYGMY-POSSUM; HERBERT RIVER RINGTAIL POSSUM; COMMON RINGTAIL POSSUM; SUGAR GLIDER; LEADBEATER'S POSSUM; EASTERN PYGMY-POSSUM; GREEN RINGTAIL POSSUM; GREATER GLIDER

TOP LEFT AND CLOCKWISE: YELLOW-BELLIED GLIDER; COMMON BRUSHTAIL POSSUM; SPOTTED CUSCUS

HUMPBACKED WHALES AT PLAY

MARINE MAMMALS

Australian coastal waters are home to three sorts of marine mammals.

Sea-lions and seals live and feed at sea.They go back onto the land to breed and to moult. Their bodies are streamlined and their forelimbs have become flippers, while their hindlimbs are used as paddles. Hunting reduced their numbers in bygone years, but they are now seen in numbers again on Australia's southern coasts.

The Dugong is a vegetarian, grazing on the seagrass beds of shallow coastal waters and estuaries. It is a gentle creature, whose survival depends on preservation of the marine pastures on which it feeds.

Dolphins and whales swim with up-and-down strokes of their horizontal tail-flukes. The Humpbacked Whale and Bottle-nosed Dolphin are well-known in Australian waters. Whales attract great interest when they pass close inshore on migration. Dolphins come to several localities around the coast, where they interact in friendly fashion with humans.

BOTTLE-NOSED DOLPHIN
OPPOSITE: AUSTRALIAN SEA-LION; INSET LEFT: BABY AUSTRALIAN SEA-LION; INSET RIGHT: AUSTRALIAN SEA-LION

MALE RED-WINGED PARROT

GANG-GANG COCKATOO

PARROTS

Parrots are birds with short, rounded bills and nostrils set in fleshy ceres. Their feathers are often colourful. Two toes on each foot point forwards and two backwards and the foot can be used for climbing, perching or holding food to be eaten. Lorikeets are a group of small parrots with tight, glossy feathers and brush-tipped tongues. They fly in noisy flocks and eat nectar and soft fruits. Cockatoos are a group of larger birds with moveable crests which are raised after alighting and to signal emotion. They eat fruit, bulbs, nuts and seed and some species also eat insects. Australia is "the land of parrots", with many spectacular and beautiful species, some common and widespread, others extremely rare.

TOP LEFT AND CLOCKWISE: PINK COCKATOO; SCALY-BREASTED LORIKEET; KING PARROT; CRIMSON ROSELLA

TOP LEFT AND CLOCKWISE: SULPHUR-CRESTED COCKATOO; YELLOW-TAILED BLACK COCKATOO; GALAHS; ECLECTUS PARROT

THE EMU'S LONG NECK ALLOWS IT A COMMANDING VIEW
OPPOSITE: EMU FEEDING; INSET LEFT: EMU; INSET RIGHT: SOUTHERN CASSOWARY

FLIGHTLESS BIRDS

Second-largest of the world's flightless birds, the Emu is unique to Australia. It may have originated as a rainforest bird, which adapted to life in open country as the continent became more arid. The Emu is widespread on the mainland except in thickly settled regions and the north-eastern forests. Individuals may wander widely. Emus eat grass, herbs and some insects. They pair during late summer and after laying on average nine or ten eggs the female wanders away, leaving the male to incubate for nearly nine weeks, then accompany the chicks for up to 18 months.

The Southern Cassowary lives in the rainforests of north-eastern Queensland. The bony casque on its head may be related to its habit of pushing through dense undergrowth with neck extended and head lowered. The female is dominant and after choosing a mate and laying four eggs she leaves the male to incubate the eggs and rear the chicks, which he defends aggressively. She may take up to three mates in succession in a season. Cassowaries feed on fallen fruit and may take animal food as well. They were much more numerous when rainforest was more extensive in Australia.

BUFF-BREASTED PARADISE-KINGFISHER

KINGFISHERS

A large bill, big head, short neck and short legs are common to all kingfishers. So is the habit of sitting motionless on an exposed perch, watching for prey to be swooped upon and battered into submission before being swallowed. Some of Australia's kingfishers feed on fish and other creatures taken in or near water, others catch insects, lizards and other small animals in woodlands and forests. The beautiful Buff-breasted Paradise-Kingfisher is a migrant to the rainforests of north eastern Queensland. Everyone's favourite kingfisher, the large Laughing Kookaburra has a pale-eyed relative, the Blue-winged Kookaburra of tropical Australia.

BLUE-WINGED KOOKABURRA
OPPOSITE: FOREST KINGFISHER

LAUGHING KOOKABURRA

WHITE-BREASTED SEA-EAGLE

OSPREY

WHISTLING KITE

BIRDS OF PREY

Australia has 24 species of birds of prey, including eagles, kites, falcons and hawks. All have powerful hooked beaks and grasping feet with sharp talons. Their eyesight is the best in the animal kingdom and all are masters of flight, adopting different hunting strategies such as hovering, stooping, stalking and explosive dashes. They hunt a variety of animal quarry ranging from grasshoppers, in the case of the Australian Kestrel, to kangaroos, in the case of the Wedge-tailed Eagle. While some take only live prey, others dispose of carrion. There is a bird of prey for every ecological niche in Australia and it is now realised that they play a vital part in controlling agricultural pests, native and introduced.

TOP LEFT AND CLOCKWISE: WEDGE-TAILED EAGLE; BLACK-BREASTED BUZZARD BREAKING EGGS WITH A STONE; AUSTRALIAN KESTREL; WEDGE-TAILED EAGLE IN FLIGHT
OPPOSITE: CRESTED HAWK AT NEST WITH CHICKS

BIRDS OF THE BUSH AND FOREST

The fairy-wrens are tiny, active birds with long, jaunty tails, restricted to Australia and Papua New Guinea. The high-pitched, reeling calls of at least one species of fairy-wren may be heard in Australia from desert to rainforest, wherever there is low vegetation. They are sociable and a nest may be attended by brilliant-plumaged breeding males, brownish females and immature birds of both sexes. Males in brown plumage can usually be told from females by their black, instead of brown, bills.

Male bowerbirds maintain areas which they decorate with leaves or with structures of sticks and other plant material. "Treasures" of some particular colour may be collected. Females attracted to the displays given in the bower areas mate, then leave to build nests and raise young. The male Satin Bowerbird of the eastern forests makes an avenue of sticks painted with plant juices, saliva and charcoal. He decorates the space at the ends of the bower with blue objects. His display includes song, mimicry and posturing with one of his treasures held in his bill.

MALE VARIEGATED FAIRY-WREN
OPPOSITE: MALE SATIN BOWERBIRD DISPLAYING TO FEMALE WHICH IS STANDING IN HIS BOWER

AUSTRALIAN PELICANS

WATERBIRDS

Australia's waterbirds may live on inland rivers, billabongs, swamps or seasonal wetlands, or around the continent's coasts. Birds which frequent fresh or brackish water include long-legged species such as egrets, herons, spoonbills, brolgas and the Jabiru. Swans, geese and ducks secure their food by grazing, diving or dabbling in shallow water. Wetland vegetation such as reeds and lilies is the habitat of crakes, waterhens and the long-toed Comb-crested Jacana. Grebes, cormorants and the Australian Darter dive for fish. Estuaries and coasts are home to residents such as oystercatchers and Australian Pelicans and a host of migrant waders which fly in for the Australian summer.

GREAT EGRET

PLUMED WHISTLING-DUCK

Australia's waterbirds are diverse in shapes, sizes and colours, but unified by grace, beauty and superb adaptation to their wetland world.

They stalk the shallows, dive for weed or fish in the deeper water, snap up minute waterlife or swallow yabbies, frogs and snakes.

In the breeding season, many put on plumes of unsurpassed elegance and greet their mates with bowing ceremony.

They are the living treasures of the estuaries, rivers, swamps and inland waters.

YELLOW-BILLED SPOONBILL; INSET, TOP LEFT AND CLOCKWISE: FRECKLED DUCK; PIED EGRET; COMB-CRESTED JACANA; BLACK SWAN AND CYGNETS

BIRDS OF THE NIGHT

Owls hunt during the night. Their soft plumage makes their flight silent and their large eyes make the best use of any light source. Acute hearing helps pinpoint prey, which is seized with powerful feet and sharp talons.

An owl's beak is narrow, short, hooked and partly hidden in feathers. Members of the barn owl group have dark eyes set in one heart-shaped disc. The other Australian owls have light eyes set in separate discs.

During daylight, roosting owls are persecuted by other birds and many wear feathers in camouflage colours. They nest in holes in trees or amongst rocks and lay white eggs. In good seasons, when prey is plentiful, the Barn Owl lays large clutches and may rear six or seven young ones.

The Tawny Frogmouth, with its wide flat bill and tiny weak feet, is not an owl at all but related to the nightjars. It eats insects and other small animals. Its nest is a flimsy platform of sticks, usually placed where a smaller branch leaves a main branch.

The Tawny Frogmouth's call is a soft but carrying, repeated "oom-oom-oom". The famous "mo-poke" call is made by the Boobook Owl. The Barn Owl makes a rasping call like cloth ripping, the Barking Owl has a deep "wook-wook" like a dog barking and the Rufous Owl calls a loud, slow "hoo-hooo".

TAWNY FROGMOUTH

SOOTY OWL

TOP LEFT AND CLOCKWISE: BARKING OWL; BARN OWL; MASKED OWL; RUFOUS OWL

ORANGE-EYED TREE-FROG

FROGS

Although frogs depend upon water in some degree or another as a habitat for their fertilised eggs and the tadpoles which hatch from them, they have adapted to a great variety of situations in Australia.

Frogs may live around the moist margins of waterways and wetlands, or in the rainforest where the air is always humid. Where the air becomes dry for part of the year or for an extended season, frogs may burrow into the ground, secrete a water-retaining covering and become torpid until rain falls again. Then they emerge and breed and their tadpoles develop quickly into adult frogs in transitory surface water.

The easiest time to locate frogs is in the breeding season, when males call loudly to advertise their presence and to attract mates. After spending some time locked together, shedding sperm and eggs, a pair of frogs separates, leaving the spawn to survive as best it may. However, the Gastric-brooding Frog nurtures its young inside its stomach.

The only toad in Australia is the introduced Cane Toad.

TOP LEFT AND CLOCKWISE: BURROWING FROG; WHITE-LIPPED TREE-FROG; GASTRIC-BROODING FROG; GIANT GREEN TREE-FROG
OPPOSITE: FLOODPLAINS FROG OF THE TOP END; INSET: MATING ORANGE-EYED TREE-FROGS

SALTWATER CROCODILE HAS BROAD SNOUT, STOUT TEETH

FRESHWATER CROCODILE HAS SLENDER SNOUT, FINE TEETH

CROCODILES

Crocodilians existed by the end of the Triassic Period (about 190 million years ago). They co-existed with the dinosaurs and the fossil records show a number of forms were still to be found in Australia when they were extinct elsewhere in the world.

Crocodiles have thick skin enclosing bony plates which serve as body armour and are well adapted to life in the water, with valves in nostrils and throat and a powerful tail for locomotion. Today, Australia is home to two representatives of this ancient family. The Saltwater Crocodile is a fierce predator which may grow to seven metres in length. The Freshwater Crocodile does not exceed three metres and eats fish and other small aquatic animals.

During the tropical Wet season, the Saltwater Crocodile lays about 60 eggs in a nest the female guards until the young hatch. The Freshwater Crocodile lays about 20 eggs in a sandbank towards the end of the Dry.

FRESHWATER CROCODILE SPRINTING INTO WATER

SALTWATER CROCODILE BASKING ON RIVERBANK

THE CROCODILE IS AN AMPHIBIOUS PREDATOR, WITH EYES AND NOSTRILS ON TOP OF ITS HEAD, A DOUBLE PALATE AND VALVES WHICH SHUT OFF NOSTRILS AND THROAT

CARPET PYTHON

SNAKES

Some lizards can be confused with snakes, but a snake has all the following characteristics: no limbs, no ear-openings, a deeply-forked tongue, eyes covered by transparent scales, a tail much shorter than the owner's snout-to-vent measurement and single underbody scales.

The 14 species of pythons found in Australia are non-venomous and crush their prey in tight body coils. They track their prey using heat-sensitive pits on their lower lips. Although they lack ears, they are very sensitive to vibration. Many other Australian snakes use venom to kill prey but are not harmful to humans. Some are potentially very dangerous to humans. Admire snakes, but if possible leave them alone!

TOP LEFT AND CLOCKWISE: DIAMOND PYTHON; EASTERN TIGER SNAKE; TAIPAN; EASTERN TIGER SNAKE EXTENDING FORKED TONGUE
OPPOSITE: GREEN TREE-SNAKE; INSET: GREEN PYTHON

EASTERN WATER DRAGON

CENTRAL NETTED DRAGON

LIZARDS

There are five different families of lizards present in Australia. Various types of dragon lizards, geckos, legless lizards, skinks and goannas come in many different shapes and sizes and occupy a wide variety of ecological niches. Like all reptiles, lizards become torpid in cold conditions and need to bask in the sun before much activity is possible. Legless lizards and some skinks may be mistaken for snakes, but have ear-openings and paired or multiple scales across their under-surfaces. There are no poisonous lizards in Australia. Most are predators, on animals ranging from termites to small mammals, though some eat plant material. Geckos, legless lizards and some skinks may sacrifice their tails to deceive an attacker.

TOP LEFT AND CLOCKWISE: PRICKLY KNOB-TAILED GECKO; THORNY DEVIL; SAND GOANNA; EASTERN WATER DRAGON; SHINGLE-BACKED LIZARD; EASTERN BLUE-TONGUE LIZARD

TOP TO BOTTOM: GIPPSLAND WATER DRAGON; VELVET GECKO; SOUTHERN FOREST DRAGON

A SPIDERWEB IS AN ELEGANT TRAP FOR INSECTS

INSECTS AND SPIDERS

Insects wear tough skeletons of chitin on the outsides of bodies which are divided into head, thorax and abdomen. The head bears a pair of sensitive antennae, while the thorax carries three pairs of legs and usually one or two pairs of wings. During the life-history of a typical insect, it undergoes one or more changes of form. For example, a butterfly lays an egg which hatches into a caterpillar or larva. This eats and sheds its skin a number of times, then becomes a chrysalis or pupa. From this resting stage the adult butterfly or imago emerges.

Insects make up the majority of the world's species of animals. Their importance in nature's cycles is inestimable. In Australia, besides devouring plants, serving as food for other animals and many other functions, they serve to pollinate many of the continent's plants.

The spider is an eight-legged arachnid. Comparison between its portrait and the insects in the other pictures will show many differences.

INSECTS SUCH AS THIS GRASSHOPPER ARE LINKS IN THE FOOD-CHAIN BETWEEN PLANTS AND PREDATORS

DRAGONFLIES HAVE AQUATIC LARVAE, AND ARE FAST-FLYING HUNTERS

TOP LEFT AND CLOCKWISE: INSECTS MAY BE INCONSPICUOUS ON THEIR NORMAL FOOD; THEY MAY BE STRIKINGLY COLOURED LIKE LEICHHARDT'S GRASSHOPPER; THE MONARCH BUTTERFLY WEARS COLOURS WARNING IT IS DISTASTEFUL; THE ST ANDREW'S CROSS SPIDER IS NOT AN INSECT; BEETLES COMPRISE A MAJORITY OF INSECT SPECIES; MOTHS GENERALLY FLY AT NIGHT

FISHES OF THE GREAT BARRIER REEF

The Great Barrier Reef covers an enormous area of the continental shelf of north-eastern Australia and supports an enormous diversity of fish - possibly more than 2,000 different species for the whole Reef. Smallest are minute gobies, which may be the tiniest animals with backbones alive. Largest are the giant predators, the sharks and rays, the marlins and rock cods. Colours and shapes range from strictly utilitarian to creatures of fantasy and beyond.

The sharks and rays have cartilaginous skeletons. They lack swim bladders and so must continue to swim or lie on the bottom. Bony fish have skeletons of bone and swim bladders filled with gas which stabilise them in the water.

Each species of fish on the Reef has its own lifestyle, its own place to live, its own way of feeding, avoiding danger and reproducing. And there are remarkable relationships here too, such as that of anemone fish and their protective hosts and cleaner fish and the species which assemble for their attentions.

TOP LEFT AND CLOCKWISE: LONG-NOSED BUTTERFLY FISH; EMPEROR ANGELFISH; GOLDEN DAMSEL; AXILSPOT HOGFISH; TRUMPETFISH; BLUESTRIPE SEAPERCH; MASKED BANNERFISH; REGAL ANGELFISH; CORAL COD AND CLEANER WRASSE

PINK ANEMONE FISH PROTECTED BY STINGING TENTACLES OF ANEMONE HOST

INVERTEBRATES OF THE GREAT BARRIER REEF

The most important "animals without backbones" on a coral reef are the reef-building coral polyps. Each intricate piece of coral is made up of thousands of individual polyps, each in the small cup of limestone it has built from material extracted from the seawater.

The Reef provides habitat for a bewildering variety of invertebrates. There are sea-jellies and anemones (both related to coral polyps), sponges, worms which display great beauty of colour and form and crustaceans, ranging from tiny shrimps to hermit-crabs and crays.

There are sea-stars, sea-urchins, sea-cucumbers and everywhere members of the mollusc family, from giant clams to cowries and poisonous cones and the intelligent, many-armed, colour-changing octopus, squid and cuttle.

TOP LEFT AND CLOCKWISE: PAINTED CRAY; CRAB GUARDING EGGS; SUNSHINE CORAL; FEATHER STAR; VERMILION BISCUIT STAR; SOFT CORAL

TINY POLYPS ARE ASSEMBLED INTO CORAL COLONIES IN A VARIETY OF FANTASTIC SHAPES AND COLOURS

  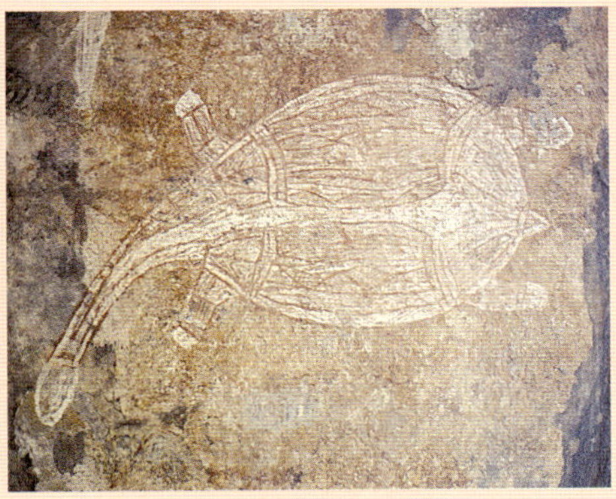

THE ABORIGINAL PEOPLE AND WILDLIFE

The Aboriginal people have lived in Australia for more than 40,000 years. They hunted the giant animals which existed until 15,000 years ago. These included the rhinoceros-like Diprotodon, a three-metre-tall kangaroo, a huge wombat and enormous flightless birds. Traditional Aboriginal stories recall these animals.

Aboriginal laws have for thousands of years governed the seasons in which certain animals could be taken and who could hunt them. These laws guarded against over-hunting and ensured suitable division of the prey.

Animals are integral parts of the Dreamtime stories and the legends of spirit ancestors and the creation of the landscape and its features.

The relationships between individual Aborigines and their totemic animals are complex.

The representation of animals on stone or bark has for thousands of years been part of Aboriginal life. Paintings, stencils and engravings exist in many places. The artistry of this work can be appreciated by most of those who view it. However, it is well to be aware that much Aboriginal art contains educational, historical or spiritual significance and sometimes all of these.

ANIMALS WHICH FIGURE IN ABORIGINAL LEGEND AND LIFE: (TOP LEFT AND CLOCKWISE): RED KANGAROO; FRILL-NECKED LIZARD; EMU; MAGPIE-GOOSE; WHISTLING-DUCKS; GREEN TURTLE

ABORIGINAL ROCK-ART, TOP LEFT AND CLOCKWISE: PIG-NOSED TURTLE; BARRAMUNDI; GOANNA; ANCESTRAL BEING AND TWO AUSTRALIAN BUSTARDS